MW00856963

IMPORTANT NOTICE

This booklet is intended to be an introductory guide, enhancing the reader's enjoyment and understanding of the flora of the coastal rainforest. I am neither a botanist nor a physician, nor am I aware of the particular physical conditions or sensitivities of any individual reader. Before attempting to use any plant as food or medicine, the reader should carefully consult additional botanical sources to confirm the identity of the plant and consult with a medical professional to ensure the use of the plant is appropriate for that reader's individual circumstances.

All plants in volume 1 grow in most or all of the coastal rainforest. Many plants listed below cross all habitats, or more than the habitat under which they are listed. In general, plants are arranged alphabetically.

Freshwater wetlands (bogs, muskegs, swamps, and wet areas):

Forest openings and open grassy meadows:

iv

"Cousin marmot, may I visit with you on this warm, blue sky summer day? . . . ."

As I wait for signs of consent to interact with my marmot friend, I pay attention to the feeling quality of my environment. I notice the sun

pleasurably warming my skin. An oxygen-drenched breeze gently blows spruce and hemlock branches, and I breathe more deeply. A pair of shimmering black ravens whoosh overhead, and my gratitude escalates. Recognizing my good feelings as consent to visit, silently I ask marmot if he has anything for me to learn today.

Again I wait, then notice marmot selectively cropping a certain kind of plant. Continuing my silent connection, I ask what he's eating, reassuring him with each careful step that I wish no harm. In close proximity now, I see that marmot eats only clover blossoms, giving the very likely indication that clover is at its nutritional best for harvest. I thank marmot for helping me learn today and wish him an abundant food supply this summer. *(Based on Project NatureConnect activities)*

Michael J. Cohen, Ed.D. provides a unique environmental education program that has progressively helped me interact and participate in an integrated way with plants, animals, birds, fish, water, sky, earth — all of nature!

Dr. Cohen is founder and Executive Director of Project NatureConnect, established in 1987 in cooperation with the Institute of Global Education—a United Nation's Non-Governmental Organization. This program provides home study and hands-on experiences in Applied Ecopsychology and Integrated Ecology.   PNC Nature-Psychology *in action*:
Home study courses, degrees, careers
www.ecopsych.com 888.285.4694

ACKNOWLEDGMENTS

My gratitude and thanks to the Life Force that provides everything, and to everyone who supported this project:

My daughter Tiffany Hirst for her faith and confidence; my father and mother Glen and Margaret Prince, my sisters Judy Hundley and Janice Prince, and my brother Bruce Prince for their significant support and individual contributions; Phil Johnson for his support, knowledge, and persistence; Nancy Karacand for her unconditional support; Jeri Museth, Stan Reddekopp, and Judy Vick for their love; Norma Strickland for her lifelong friendship and business sense; Judy Jones for her friendship and editing contributions; Nora Laughlin for her friendship and generosity; Bridget Riley for her motivation.

Marie Olson of the Eagle Moiety Wooshkeetaan Clan of Auke Bay, Alaska, provided Tlingit names and usage for a number of plants in this booklet. With permission, I am including available Tlingit plant names in recognition and honor of indigenous cultures everywhere who have given the world valuable knowledge.

INTRODUCTION

Staring into the beautiful, sunlit face of a radiantly vibrant pink flower, I ask if I may eat it and use its salmonberry energy today. Listening quietly for a response, I sense its friendly essence reaching out, giving me a warm hug—giving permission. Sweetness flows through me as I eat the brilliant blossom, and I am thankful for its gift.

For over 23 years I have been learning about edible and medicinal properties of wild plants. Using ingredients directly from nature provides satisfaction at many different levels. Profound insights and healing moments occur while gathering edible plants. For instance, while collecting nettles one sunny spring morning, I understood that everything in nature is interconnected. Spruce and hemlock, ravens and eagles, salmon and whales, bees and mosquitoes, fireweed and salmonberries, bears and people, we all use the same earth, air, sun, and water.

Made of the same basic elements, we are all related.

Common Name: **SITKA BURNET**
Scientific Name: *Sanguisorba sitchensis*
Plant Family: Rosaceae "The Rose Family"
Habitat: Bogs, swamps, moist meadows, stream banks
Edible and/or medicinal use:

On a sparkling summer afternoon Sitka Burnet's delicate white flower attracted my attention, and that began my identification of this member of the rose family.

Sitka burnet leaves can be used in spring salads. In fact, another common name for this plant is salad burnet.

Leaves may also be used in herbal teas. Preparations made from roots have been used to treat dysentery, stop external bleeding, and as an antibacterial.

Common Name:	**CLOUDBERRY**
Tlingit Name:	**néx̱ˈw**
Scientific Name:	*Rubus chamaemorus*
Plant Family:	Rosaceae "The Rose Family"
Habitat:	Muskegs, wet forest openings

Edible and/or medicinal use:

Fruit of the cloudberry is one of my more recent discoveries. Previously, I have seen only their large white blossoms speckling boggy areas. During their ripe season, usually I have been picking blueberries or red huckleberries.

Ripe cloudberries taste spicy, reminiscent of cloves, and can be cooked in jelly and jam. Their autumn leaves add spectacular reds and gold to muskeg areas.

Common Name:	**BOG CRANBERRY**
Tlingit Name:	**Dáxw**
Scientific Name:	*Oxycoccus microcarpus*
Plant Family:	Ericaceae "The Heath Family"
Habitat:	Muskegs

Edible and/or medicinal use:

Bog cranberries compare in flavor to commercial cranberries. Besides their traditional use as relish with turkey or wild game, they make good jams and jellies and mix well with other berries. As with lingonberries, they are best picked after the first frost.

Medicinally cranberry juice is used for colds and bleeding gums, as a drink for urinary tract infections, and in concentrated pulp diluted in water to help asthmatics overcome an attack.

Common Name:	**LABRADOR TEA**
Tlingit Name:	**s'ikshaldéen**
Other Names:	Hudson's Bay tea
Scientific Name:	*Ledum groenlandicum*
Plant Family:	Ericaceae "The Heath Family"
Habitat:	Muskegs, bogs

Edible and/or medicinal use:

Tlingits kept Labrador tea, mixed with stewed seedless rosehips, brewing on the stove year-round, adding fresh leaves and water regularly. This brew was used to combat sore throats.

Moderate use has been recommended because of tannic acid.

Labrador tea has been confused with **poisonous bog laurel** making positive plant identification essential.

Common Name:	**LINGONBERRY**
Other Names:	lowbush cranberry, mountain cranberry
Scientific Name:	*Vaccinium vitis-idaea*
Plant Family:	Ericaceae "The Heath Family"
Habitat:	Muskegs, open woods

Edible and/or medicinal use:

Very tangy raw, lingonberries are delicious when sweetened and cooked. Easily identified by black dots on leaf undersides, they are best picked after the first fall frost. High in pectin content for use in jellies, they can be used also in baked goods, and as juice, syrup, wine, sauce, catsup, and tea.

As tea, they aid digestion by stimulating gastric juices. Leaves can be used in herbal steams and rinses because of their astringent and antiseptic properties.

Common Name:	**SKUNK CABBAGE**
Tlingit Name:	**x'áal'**
Other Names:	swamp cabbage, Alaska crocus
Scientific Name:	*Lysichiton americanum*
Plant Family:	Araceae "The Arum Family"
Habitat:	Swamps, muskeg, wet forests, meadows

Edible and/or medicinal use:

> The Tlingit baked salmon wrapped in skunk cabbage leaves in earthen fire pits. Cooked in this manner the leaves moisturized the salmon so that even the nutritious bones became soft enough to eat.

> Tea made from dried roots has been used medicinally to relieve cramps, coughs, bronchitis, asthma and inflammation of mucous membranes.

Common Name:	**ROUND-LEAVED SUNDEW**
Scientific Name:	*Drosera rotundifolia*
Plant Family:	Droseraceae "The Sundew Family"
Habitat:	Muskeg meadows, bogs

Edible and/or medicinal use:

While lying on the wet muskeg one cool foggy day, I watched the sticky tentacles of a sundew plant capture an insect in slow motion. They capture and digest many of the insects that pollinate them.

Sundew in tincture form has been used to relieve cramps, promote mucous discharge, treat asthma, bronchitis, whooping cough, and other respiratory problems.

Common Name:	**CHOCOLATE LILY**
Tlingit Name:	**kóox**
Other Names:	Northern rice root, black lily
Scientific Name:	*Fritillaria camschatcensis*
Plant Family:	Liliaceae "The Lily Family"
Habitat:	Meadows, moist open places

Edible and/or medicinal use:

Chocolate lily bulbs look like tight clusters of white rice and have been an important food to many Northwest and Alaska indigenous groups.

The Tlingit word later came to include the same rice one finds in grocery stores. The bulb, which was picked at any time of the year, was soaked in water to remove bitterness and then mixed with other berries. They may be used in stir-fries and mixed with other vegetables, as well.

Common Name:	**DWARF DOGWOOD**
Tlingit Name:	**k'eikaxetl'k**
Other Names:	bunchberry (more common), Jacob berry
Scientific Name:	*Cornus canadensis*
Plant Family:	Cornaceae "The Dogwood Family"
Habitat:	Spruce and cedar forests, meadows, bogs

Edible and/or medicinal use:

Dwarf dogwood is known more commonly as bunchberry, though I prefer dwarf dogwood because of its resemblance to the ornamental flowering dogwood. After removing seeds from these berries the Tlingit used them to thicken a mixture of blueberries, salmonberries, and thimbleberries.

Late last summer I watched a young squirrel eating its way through a patch of dwarf dogwood berries, spitting out the seeds like a peashooter, thus helping the plant enlarge its territory.

Common Name: **FIREWEED (pictured at right)**
                                DWARF FIREWEED, or River Beauty (at left)

Tlingit Name: **lóol**

Scientific Name: *Epilobium angustifolium*
                                Epilobium latifolium

Plant Family: Onagraceae "The Evening Primrose Family"

Habitat: Burned or logged areas, meadows, roadsides

Edible and/or medicinal use:

All parts of this brilliant stately plant may be eaten in a variety of ways at different times of the year. Flowers and leaves blend well with other wild berry leaves and herbs for superb tea. Fireweed honey further enhances this delightful tea.

Medicinally tea may provide relief for constipation (though too much may have a laxative effect), upset stomach, and asthma, and serves as a wonderful spring tonic.

Common Name:	**GERANIUM**
Scientific Name:	*Geranium erianthum*
Plant Family:	Geraniaceae "The Geranium Family"
Habitat:	Moist open forests, meadows, roadsides to above timberline

Edible and/or medicinal use:

Steamed or cooked, leaves may be added to soups, stews or casseroles. Lovely lavender geranium flowers offer an attractive garnish for salads. Tea from the leaves may soothe sore throats. The astringent properties of roots have been useful in treating hemorrhoids, bleeding piles, inflamed diverticula, and vaginitis.

Prior to the flower stage, geranium may be confused with **deadly poisonous monkshood** because of a resemblance in leaf structure.

Common Name:	**GOLDENROD**
Scientific Name:	*Solidago canadensis*
Plant Family:	Asteraceae "The Sunflower Family"
Habitat:	Meadows, roadsides, forest openings, disturbed areas

Edible and/or medicinal use:

Deep yellow goldenrod flowers brighten my summer walks immensely, their sunny brilliance drawing my attention long before I reach them. Flowers mix well in herbal tea blends, and greens can be added to soups, stews and salads.

Goldenrod is considered astringent, diuretic, beneficial for the kidneys, helpful in passing kidney stones, and has been used as an antiseptic wash for wounds, and treating insect bites.

Common Name: **FALSE HELLEBORE**
Scientific Name: *Veratrum viride*
Plant Family: Liliaceae "The Lily Family"
Habitat: Moist meadows, moist forest openings

NO EDIBLE USE

****EXTREMELY POISONOUS****

This plant can be confused with twisted stalk and other young edible spring greens.

While only two poisonous plants are pictured in this booklet, many toxic or deadly plants grow in the rainforest. It is literally a matter of life and death to learn plants thoroughly before attempting to use them as food or medicine.

Common Name:	**HIGHBUSH CRANBERRY**
Tlingit Name:	**kaxwéix**
Scientific Name:	*Viburnum edule*
Plant Family:	Caprifoliaceae "The Honeysuckle Family"
Habitat:	Moist forests, forest edges, riverbanks

Edible and/or medicinal use:

Traditionally highbush cranberry has been used as a sauce for wild and domestic meats and can be made into tasty catsup.

Tlingit women mashed and mixed loose salmon eggs with highbush cranberries as one food source. Elderberries mixed with highbush cranberries provided another food combination.

Medicinally the bark has been useful in treating menstrual cramps, stomach cramps, asthma, strained muscles, colds and upset stomachs.

Common Name:	**NAGOONBERRY**
Tlingit Name:	**neigóon**
Scientific Name:	*Rubus arcticus*
Plant Family:	Rosaceae "The Rose Family"
Habitat:	Moist meadows, stream banks

Edible and/or medicinal use:

The nagoonberry, one of the world's best tasting berries, makes beautiful, clear, ruby red juice, or wine for Thanksgiving and Christmas holidays for those who can take alcoholic beverages. As with most other berries, they can be used in many creative ways besides the traditional jams, jellies, pies, baked goods, juices, and yogurt.

Discovering a patch of ripe nagoonberries causes hikers long delays on sunny summer days! Mmmm Mmmmm!

Common Name:	**STINGING NETTLE**
Scientific Name:	*Urtica dioica*
Plant Family:	Urticaceae "The Nettle Family"
Habitat:	Moist rich soil, forest openings, stream banks

Edible and/or medicinal use:

One of my favorite herbs, nettles act as a restorative spring tonic spiritually, physically, mentally and emotionally. They cleanse the intestinal tract from heavier winter food. All parts of the plant may be used but should be steamed, dried or cooked before eating in order to dissolve the stingers. Cooking water from nettles makes good soup stock.

Nettles, with their high chlorophyll content, are considered an excellent blood tonic. Cramp relieving and diuretic, they may help with asthma, hay fever and prostate problems, as well.

Common Name:	**ROSE**
Tlingit Name:	**k'inchéiyi** (rosehips)
Scientific Name:	*Rosa sp.*
Plant Family:	Rosaceae "The Rose Family"
Habitat:	Meadows, forests, roadsides, mountain slopes

Edible and/or medicinal use:

Rose petals add delicate flavor in salads, jellies, honey, syrup, and drinks. Many commercial herb teas contain rosehips because of their high vitamin C content.

As an external wash, rose petal tea relieves dry skin; inhaling rose oil vapor soothes headaches, nausea and insomnia; rose honey alleviates sore throat. One winter, following an Adelle Davis recipe, we made and drank many quarts of rosehip concentrate in various liquid concoctions, and enjoyed the winter without a trace of colds or flu.

Common Name:	**SALMONBERRY**
Tlingit Name:	**wasˈxˈaan tléigu**
Scientific Name:	*Rubus spectabilis*
Plant Family:	Rosaceae "The Rose Family"
Habitat:	Moist woods, stream edges, low mountain regions

Edible and/or medicinal use:

Related to the raspberry, fresh salmonberry sprouts, flowers, leaves and berries are edible. Flowers (pictured on back cover) and leaves may be dried for tea. Salmonberries may be red or orange and mix wonderfully with other berries. They may be added to baked goods, cereals, puddings, and used in juices or alcoholic beverages. I find these berries bland and prefer them mixed with more flavorful berries. Alone or in salads, blossoms add sweetness and visual pleasure.

Common Name:	**THIMBLEBERRY**
Tlingit Name:	**ch'ee<u>x</u>' or ch' ei<u>x</u>'** (depending on sentence)
Scientific Name:	*Rubus parviflorus*
Plant Family:	Rosaceae "The Rose Family"
Habitat:	Forest openings, open areas, roadsides

Edible and/or medicinal use:

> Velvety red thimbleberry is another raspberry relative. Large leaves and brilliant white flowers may be dried for teas, though if leaves are used for tea, they **must** be dried thoroughly. Berries may be mixed with other more flavorful berries in a variety of ways.

> Thimbleberry root has been used for treating diarrhea.

Common Name:	**TWISTED STALK**
Tlingit Name:	**tlei<u>k</u>w kahínti**
Other Names:	watermelon berry
Scientific Name:	*Streptopus amplexifolius*
Plant Family:	Liliaceae "The Lily Family"
Habitat:	Moist woods, open areas, stream banks

Edible and/or medicinal use:

Delicious in spring, tasting like cultivated cucumber, twisted stalk can be peeled and eaten raw, steamed or stir-fried with other vegetables, or added to soups, casseroles and salads. They are one of my favorite trail foods after the snow first melts in early spring. Berries taste like watermelon.

NOTE: Young spring shoots can be confused with **deadly false hellebore**.

Common Name: **YELLOW RATTLE**
Scientific Name: *Rhinanthus minor*
Plant Family: Scrophulariaceae "The Figwort Family"
Habitat: Meadows, open grassy areas, roadsides, moist slopes
Edible and/or medicinal use:

>Yellow rattle has been used to treat children's pinkeye. It is considered a substitute for eyebright, another medicinal plant used as an eyewash.

>According to Janice Schofield in <u>Discovering Wild Plants,</u> yellow rattle is a semiparasitic plant that wraps its roots around roots of grass, attaching nodules that drain nutrients and water intended for the grasses.

Common Name:	**BANEBERRY**
Scientific Name:	*Actaea rubra*
Plant Family:	Ranunculaceae "The Buttercup Family"
Habitat:	Moist, shady forest, open woods, stream banks

NO EDIBLE USE:

****POISON AND DEADLY****

All parts of the plant are poisonous.

Berries can be RED or WHITE

Common Name:	**OREGON CRAB APPLE**
Tlingit Name:	**lingít x'áax'i**
Other Names:	Pacific crab apple
Scientific Name:	*Malus fusca*
Plant Family:	Rosaceae "The Rose Family"
Habitat:	Moist woods, swamp edges

Edible and/or medicinal use:

Ripe in early fall and very tart, this fruit provides a good source of pectin for making jelly and mixing with other kinds of fruit. It has been an important food to indigenous people.

The bark has been used medicinally, though only by those knowledgeable of its properties since it contains toxic compounds.

Common Name:	**BRISTLY BLACK CURRANT**
Tlingit Name:	**x̲aaheiwú**
Scientific Name:	*Ribes lacustre*
Plant Family:	Saxifragaceae "The Saxifrage Family"
Habitat:	Moist woods, stream banks, seashore

Edible and/or medicinal use:

Fruits and stems may be either prickly or smooth. Currants mix well with other berries in jellies, pies, alcoholic beverages, syrup, and juices.

Exceptionally high in vitamin C and copper, they stimulate the appetite, and help heal mouth inflammation and sore throat. Tea made from leaves may be useful in treating rheumatic problems. Currant seed oils have been used in treatment of alcoholism.

Common Name:	**DEVIL'S CLUB**
Tlingit Name:	**s'axt**
Scientific Name:	*Oplopanax horridum*
Plant Family:	Araliaceae "The Ginseng Family"
Habitat:	Moist woods

Edible and/or medicinal use:

Richly resinous scented new growth is safe to eat for a few days after they first appear, and only a few buds are recommended to add flavor and nutrition to soups or other edible preparations. Thereafter, tea from roots collected in spring provides a general body-balancing tonic.

Part of the Oriental ginseng family, devil's club has been reported to lower blood sugar levels, so caution is required for anyone with diabetes or other blood sugar related problems.

Common Name:	**RED ELDERBERRY**
Tlingit Name:	**yéil**[|]
Scientific Name:	*Sambucus racemosa*
Plant Family:	Caprifoliaceae "The Honeysuckle Family"
Habitat:	Forest openings, stream banks, roadsides

Edible and/or medicinal use:

Elderberry seeds are POISONOUS WHEN RAW, but the pulp from the cooked fruit, after removing seeds, can be used in jellies, herb vinegar, wine, syrup and juice.

Elder flowers have been helpful in treating rheumatism, constipation, colds, flu, sinus problems, and fevers and are known for their sedative properties as well.

Common Name:	**RED HUCKLEBERRY**
Tlingit Name:	**tleikatánk**
Scientific Name:	*Vaccinium parvifolium*
Plant Family:	Ericaceae "The Heath Family"
Habitat:	Forest edges, forest openings

Edible and/or medicinal use:

Beautiful red huckleberries make one of the best tasting pies! As with its blueberry relatives, huckleberries offer wide versatility in baked goods, jam, jelly, wine, juice, ice cream, yogurt, crepes, syrup, and any other creative way you can imagine.

They may help maintain stable blood sugar and stimulate appetite.

Common Name:	**JEWELWEED**
Scientific Name:	*Impatiens noli-tangere*
Plant Family:	Balsaminaceae "Touch-Me-Not Family"
Habitat:	Stream banks, moist forests

Edible and/or medicinal use:

In order to qualify as edible for human consumption, some plants must be collected before a certain stage and cooked first. Jewelweed is such a plant. It can be eaten only while still a young shoot less than a foot high, and it must be cooked in more than one change of boiling water before being considered an edible vegetable.

Jewelweed has been used to treat hives, and to prevent rashes from poison ivy, stinging nettles, and other rash-causing plants.

Common Name:	**MONKEY FLOWER**
Scientific Name:	*Mimulus guttatus*
Plant Family:	Scrophulariaceae "The Figwort Family"
Habitat:	Drainage ditches, stream edges, wet areas

Edible and/or medicinal use:

Sunny yellow monkey flowers often mix with blue forget-me-nots in wet ditches along roadsides. Going for a walk on a road lined with this lovely natural attraction always brightens my spirit.

Stems and leaves can be added to salads and soups, and steamed with other vegetables. Blossoms are edible too.

Leaves and stems can be used as a poultice for cuts, scratches, and insect bites.

Common Name:	**TRAILING RASPBERRY**
Other Names:	Five-leaved bramble
Scientific Name:	*Rubus pedatus*
Plant Family:	Rosaceae "The Rose Family"
Habitat:	Moist forest, stream banks

Edible and/or medicinal use:

Difficult to collect in quantity because of its small size, trailing raspberry is probably enjoyed better while on the hiking trail. Its rich green leaves and tiny ruby-red berries contrast strikingly against the paler green moss in which it is often found in old growth forests.

While the trailing raspberry pictured appears large, it is actually a small plant. Showing close up detail can lead to deceptive impressions of a particular plant, making identification from a number of references significantly important.

Common Name:	**BEACH GREENS**
Other Names:	Seabeach sandwort, sea chickweed
Scientific Name:	*Honckenya peploides*
Plant Family:	Caryophyllaceae "The Pink Family"
Habitat:	Sandy, rocky beaches

Edible and/or medicinal use:

Young shoots and leaves of this bright green beach plant add variety and a touch of the ocean to salads. They can be steamed lightly or included in stir-fry dishes.

Beach greens are high in vitamins A and C.

Common Name: **BEACH PEA**
Scientific Name: *Lathyrus maritimus*
Plant Family: Fabaceae "The Pea Family"
Habitat: Coastal areas along sandy or gravel beaches
Edible and/or medicinal use:

> Beach peas are another plant around which caution needs to be used. Unless you are sure of the identity of the species, don't eat it. Some of the pea family contain toxins which can cause temporary or permanent paralysis. The edible variety can be eaten in the same manner as the cultivated garden variety, such as in stir-fries and salads, though moderation is advised.
>
> Medicinally they can be helpful to the health of the intestinal tract.

Common Name:	**GOOSETONGUE**
Tlingit Name:	**suktéitl**[1]
Other Names:	sea plantain
Scientific Name:	*Plantago maritima*
Plant Family:	Plantaginaceae "The Plantain Family"
Habitat:	Salt marshes, sandy or gravelly beaches

Edible and/or medicinal use:

One of my favorite wild edibles, goosetongue steamed lightly tastes much like asparagus. Edible raw, it can be blanched and frozen or jarred as well. It tastes great right off the beach, lightly steamed, or added to salads and stir-fries.

Common Name:	**ROSEROOT**
Scientific Name:	*Sedum rosea*
Plant Family:	Crassulaceae "The Stonecrop Family"
Habitat:	Moist rocky slopes and cliffs

Edible and/or medicinal use:

> Best collected before they flower, the succulent leaves are edible raw, or steamed as greens. The rootstock collected in spring or fall provides another food source. Cut or bruised roots of this plant emit a delicate rose scent.

> Indigenous groups used this plant to treat sore throats, colds, and as an eyewash.

Common Name: **FIELD HORSETAIL**
Scientific Name: *Equisetum arvense*
Plant Family: Equisetaceae "The Horsetail Family"
Habitat: Moist to wet forests, meadows, waste areas, marshes
Edible and/or medicinal use:

Prehistoric horsetail, one of the oldest plants living today, is also one of the most widespread plants in the world—a marvelous survivor! Besides edible and medicinal properties, ancient and modern people alike have used it as a cleanser and polisher.

Cooked young horsetail greens can be used as a vegetable. They must be cooked. Part of the root can be eaten in early spring.

Medicinally horsetail has been used as a poultice for growths and hemorrhages. As a tea it has been useful in treating kidney stones, rheumatism, bladder and urinary tract diseases, and stomach ulcers.

Common Name: **PINEAPPLE WEED**
Scientific Name: *Matricaria matricarioides*
Plant Family: Asteraceae "The Sunflower Family"
Habitat: Roadsides, disturbed soils, waste areas
Edible and/or medicinal use:

Flower heads of pineapple weed add a fragrant pineapple scent to salads and other vegetable dishes, along with a fresh, pleasant taste.

As a medicinal preparation, tea made from the flowers of this plant may provide treatment for insomnia or nervousness, gas and upset stomach, and it acts as an anti-inflammatory, as well.

Common Name:	**YARROW**
Scientific Name:	*Achillea borealis*
Plant Family:	Asteraceae "The Sunflower Family"
Habitat:	Dry to moist meadows, open areas, roadsides

Edible and/or medicinal use:

Yarrow, an herb with many attributes, has been used to treat fevers, colds, sore throats, flu, irregular menstruation, menopause, for its effect on the circulatory, digestive, and urinary systems, and as a blood renewal stimulant. Yarrow has been known to break up a cold within 24 hours.

Growing almost everywhere, flowers may be white to deep pink or reddish in our rainforest.

References

Cohen, Michael J. (1997). <u>Reconnecting With Nature</u>. Corvallis: Ecopress.

Cooperative Extension Service. (1976). <u>Wild Edible and Poisonous Plants of Alaska</u>. College: University of Alaska.

Keville, Kathi. (1996). <u>Herbs for Health and Healing</u>. Emmaus: Rodale Press, Inc.

Ody, Penelope. (1993). <u>The Complete Medicinal Herbal</u>. New York: DK Publishing, Inc.

Pojar, Jim and Andy MacKinnon, eds. (1994). <u>Plants of the Pacific Northwest Coast, Washington, Oregon, British Columbia & Alaska</u>. Vancouver: Lone Pine Publishing.

Schofield, Janice J. (1989). <u>Discovering Wild Plants, Alaska, Western Canada, The Northwest</u>. Anchorage: Alaska Northwest Books.

Wren, R.W., ed. (1972). <u>Potter's New Cyclopaedia of Medicinal Herbs and Preparations</u>. New York: Harper & Row, Publishers.

Interview:

Olson, Marie. (1997). Personal Interview. <u>Conversation with Tlingit elder of The Alaska Native Sisterhood, Camp 2</u>. Juneau, Alaska.

Telephone and Written Interviews:

Stensvold, Mary. (1998). Personal Interviews and Written Communications. <u>Conversations with Mary Stensvold (Regional Botanist, USDA Forest Service, Alaska Region, Sitka, Alaska) verifying scientific and plant family names and species identification</u>.

INDEX